行走在黄冈大别山世界地质公园

Walking in Huanggang Dabieshan UNESCO Global Geopark

李江风　彭文胜　黎　妍　邓志高　高志峰　吴　笛　编

张地珂　李　凤　杜冬琴　译

中国地质大学出版社
CHINA UNIVERSITY OF GEOSCIENCES PRESS

目录 Contents

引言

Introduction

十二年磨一剑 / Twelve years' Development /

故事要从2006年大别山的秋天说起。

作为中国中央山系地质-地理-生态-气候分界线的重要组成部分，黄冈大别山具有极其重要的地位，它拥有地质遗迹的典型性、完整性、系统性、稀有性和优美性，生态环境优良、历史文化厚重、科普价值极高，孕育了大别山世代人民。而大别山土生土长的人们，一直以来都有守护家园、保护大别山的心愿，这个心愿在不断的经济发展和社会进步中悄然发芽。于是，在2006年9月，黄冈市人民政府正式提出了申报大别山国家地质公园的构想，申报工作从此拉开了序幕。

经过长达12年的漫长努力，黄冈大别山从建设成为省级地质公园、国家地质公园，到2018年4月17日，正式建设成为黄冈大别山联合国教科文组织世界地质公园。公园的创建之路，充满了艰辛。它的建成，在黄冈发展史上具有里程碑意义，开启了黄冈大别山地质资源保护与发展的新篇章。

The story of Huanggang Dabieshan UNESCO Global Geopark began in the autumn of 2006.

As an important dividing line of geography, geology, climate and ecology of the Central Moutains in China, Huanggang Dabieshan UNESCO Global Geopark is of great importance in location. It comprises a number of geological heritage sites of special scientific importance,

integrity, rarity, systematicness and beauty. It is a nature protected area with good ecological environment, profound culture and high scientific value that has given birth to its Dabieshan people generation after generation who have always had a wish to protect their hometown — Dabieshan. With the development of economy and social progress, this wish grows quietly and instantly. Therefore, in September 2006, the People's Government of Huanggang City officially submitted the proposal for a UNESCO Global Geopark.

After 12 years of hard work, Huanggang Dabieshan grows from a provincial geopark to a national geopark, and finally to Huanggang Dabieshan UNESCO Global Geopark on April 17, 2018. Full of hardship, the completion of the park is a millstone in the protection and development of geological resources of Dabieshan Mountains.

时　间	公园大事记
2006年9月	黄冈市人民政府提出了申报大别山国家地质公园的构想
2006年9月22日	黄冈市人民政府启动省级地质公园申报工作
2007年3月29日	湖北大别山（黄冈）地质公园被批准为省级地质公园，同时启动国家地质公园申报工作
2009年8月19日	湖北大别山（黄冈）地质公园被国土资源部批准为第五批国家地质公园建设单位
2012年11月14日	湖北黄冈大别山国家地质公园通过国土资源部专家组的验收
2012年12月25日	湖北大别山（黄冈）地质公园被国土资源部正式命名为湖北黄冈大别山国家地质公园
2013年8月5日	公园被命名为全国第三批“国土资源科普基地”
2013年10月27日	入选全国首届“中国最美地质公园”前30名
2013年11月2日	在国土资源部地质环境司的主持下，国家地质公园正式揭碑开园
2013年11月25日	公园被湖北省科学技术协会命名为“湖北省科普教育基地”
2015年12月27日	在全国第九批世界地质公园推荐评审会上，湖北黄冈大别山国家地质公园以第一名的成绩，获得全国2017年度世界地质公园申报资格
2017年7月15日	黄冈大别山地质公园博物馆举行新馆开馆仪式，并作为中国地质博物馆黄冈分馆正式对外开放
2017年7月17日至20日	公园接受联合国教科文组织世界地质公园评估专家克劳斯·乔治和全勇文的实地考察评估
2018年4月17日	联合国教科文组织世界地质公园理事会第204次会议通过决议：批准湖北黄冈大别山地质公园成为联合国教科文组织世界地质公园
2018年9月13日	第八届世界地质公园大会上，中国黄冈大别山世界地质公园被正式授牌为世界地质公园网络成员
2018年11月7日	在2018中国联合国教科文组织世界地质公园年会上，国家林业和草原局自然保护地管理司司长杨超为黄冈大别山世界地质公园授牌
2019年5月9日至12日	黄冈大别山世界地质公园举办揭碑开园暨地质公园与区域经济发展国际研讨会

大别山画卷 Dabieshan picture scroll
摄影：刘志雄 Photographed by Liu Zhixiong

公园位置 / Location /

黄冈大别山联合国教科文组织世界地质公园，总面积为2 625.54km²，位于中华人民共和国湖北省黄冈市境内，介于东经115° 03′ 13″ —115° 52′ 18″ 、北纬30° 43′ 46″ —31° 17′ 18″ 之间，海拔范围314～1 729.13m，地跨麻城市、罗田县和英山县。该园属长江中下游亚热带季风气候区，以大陆造山带构造-花岗岩山岳地貌为主，地质遗迹及生物资源丰富，历史文化悠久。

Huanggang Dabieshan UNESCO Global Geopark, with an area of 2 625.54km², is located in Huanggang City, Hubei Province, People's Republic of China, of which the geographic coordinates is at 115°03'13"-115°52"18"E, and 30°43'46"-31°17'18"N and its altitude ranges from 314m to 1 729.13m. It is in the subtropical monsoon climate zone of the middle and lower reaches of Yangtze River and straddles Macheng, Luotian and Yingshan. Mainly characterized by structure landform and granite landform of the continental orogenic belt, the Geopark boasts spectacular geological sites, abundant biological resources, and profound history and culture.

黄冈大别山世界地质公园主碑
Main Monument of Huanggang Dabieshan UNESCO Global Geopark

黄冈大别山世界地质公园Logo
Logo of Huanggang Dabieshan UNESCO Global Geopark

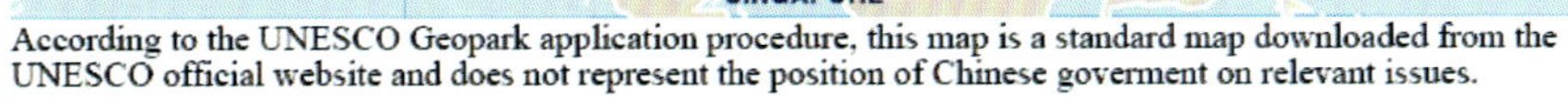
According to the UNESCO Geopark application procedure, this map is a standard map downloaded from the UNESCO official website and does not represent the position of Chinese goverment on relevant issues.

黄冈大别山世界地质公园在东亚的位置
Location of Huanggang Dabieshan UNESCO Global Geopark in East Asia

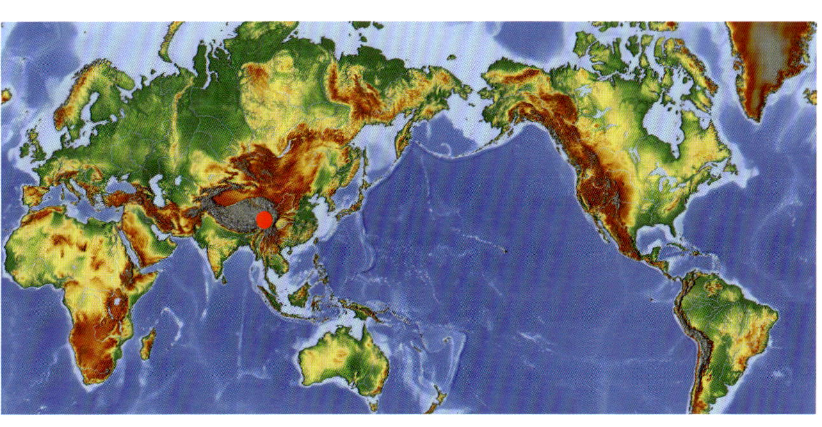

黄冈大别山地质公园在全球造山带上的位置
Location of Huanggang Dabieshan UNESCO Global Geopark on the Global Orogenic Belt

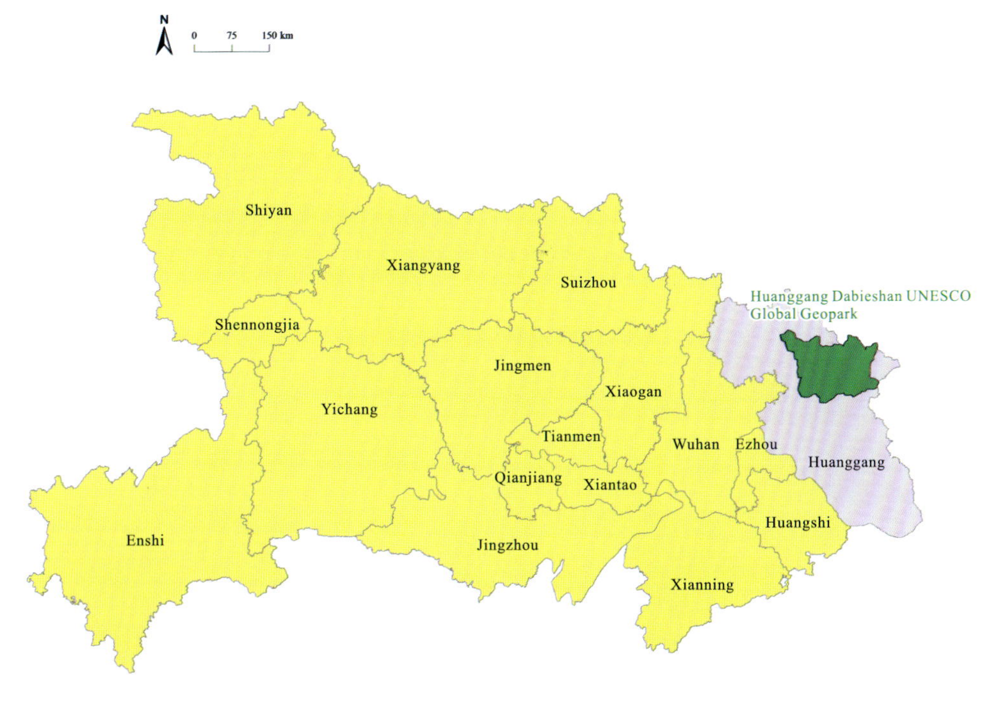

黄冈大别山世界地质公园在中国湖北的位置
Location of Huanggang Dabieshan UNESCO Global Geopark in Hubei Province, China

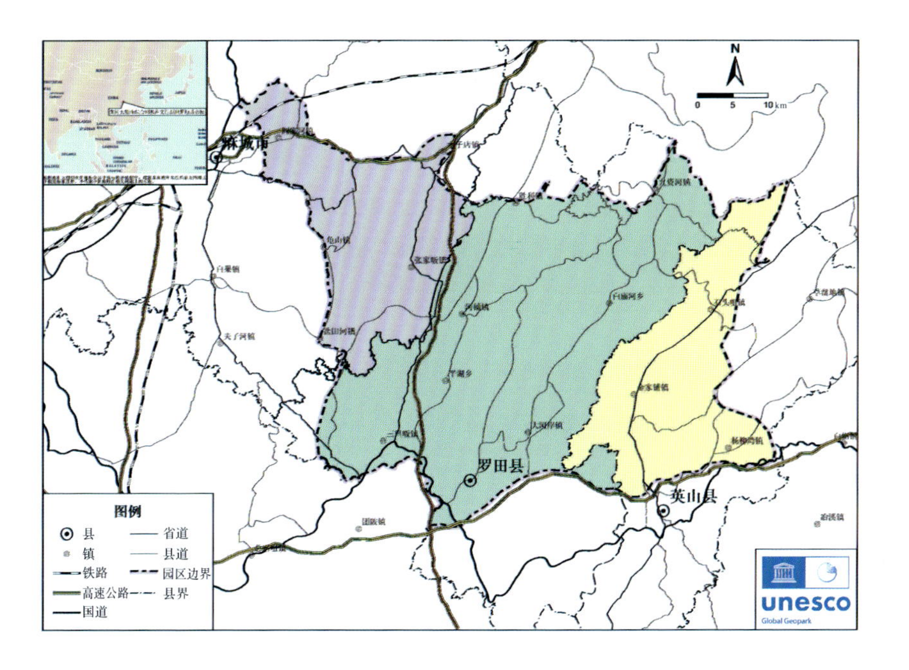

黄冈大别山世界地质公园范围图
Map of Huanggang Dabieshan UNESCO Global Geopark

公园地质遗迹点 / Geosites /

公园内重要的地质遗迹点有新太古代麻粒岩相片麻岩、新太古代TTG岩系、朱家河高压榴辉岩、陶家山超高压榴辉岩、新太古代木子店组古陆核、三合湾韧性剪切带、中侏罗世片麻状花岗岩和白垩系红层，早白垩世、晚白垩世花岗岩及其地貌景观，以及神侣沟、龙潭河谷等流水侵蚀地貌景观和天堂崖、百丈崖、云崖瀑布等水体景观。

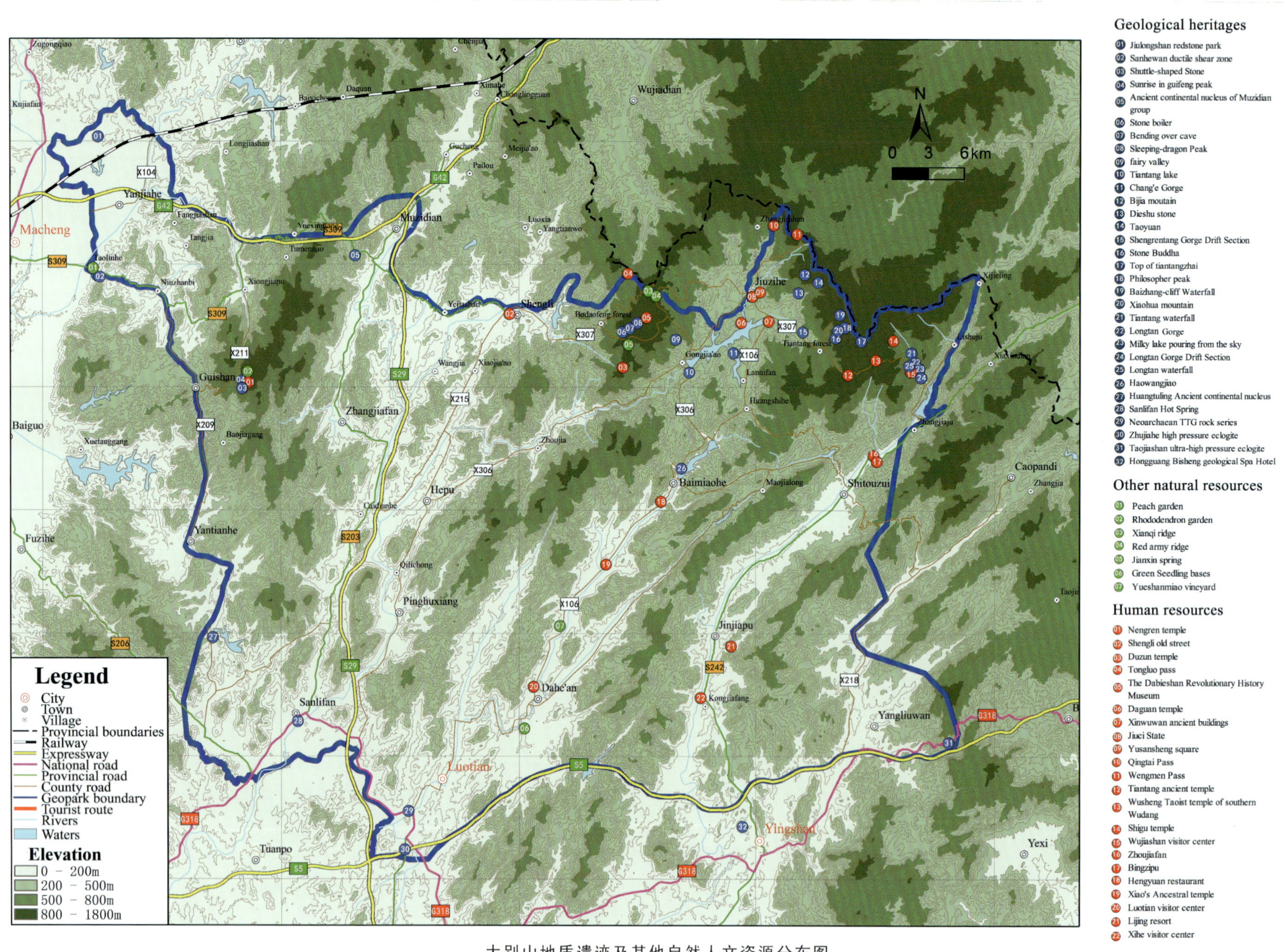

大别山地质遗迹及其他自然人文资源分布图
Distribution map of geological relics and other natural and cultural resources in Dabieshan

There are a number of important geosites in the Geopark, such as Neoarchean granulite-facies gneiss, Neoarchean Trondhjemite-Tonalite-Granodiorite, Zhujiahe high-pressure eclogite, Taojiashan ultra-high pressure eclogite, Neoarchean Muzidian Formation ancient continental nucleus, Sanhewan ductile shear zone; Middle Jurassic gneissic granite and Cretaceous red-bed; Early Cretaceous and Late Cretaceous granite and its landscape; and some fluvial-erosion landforms, such as Shenlvgou gorge and Longtan gorge etc., and some water landscapes, for example, Paradise Waterfall, Baizhangya Waterfall and Yunya Waterfall.

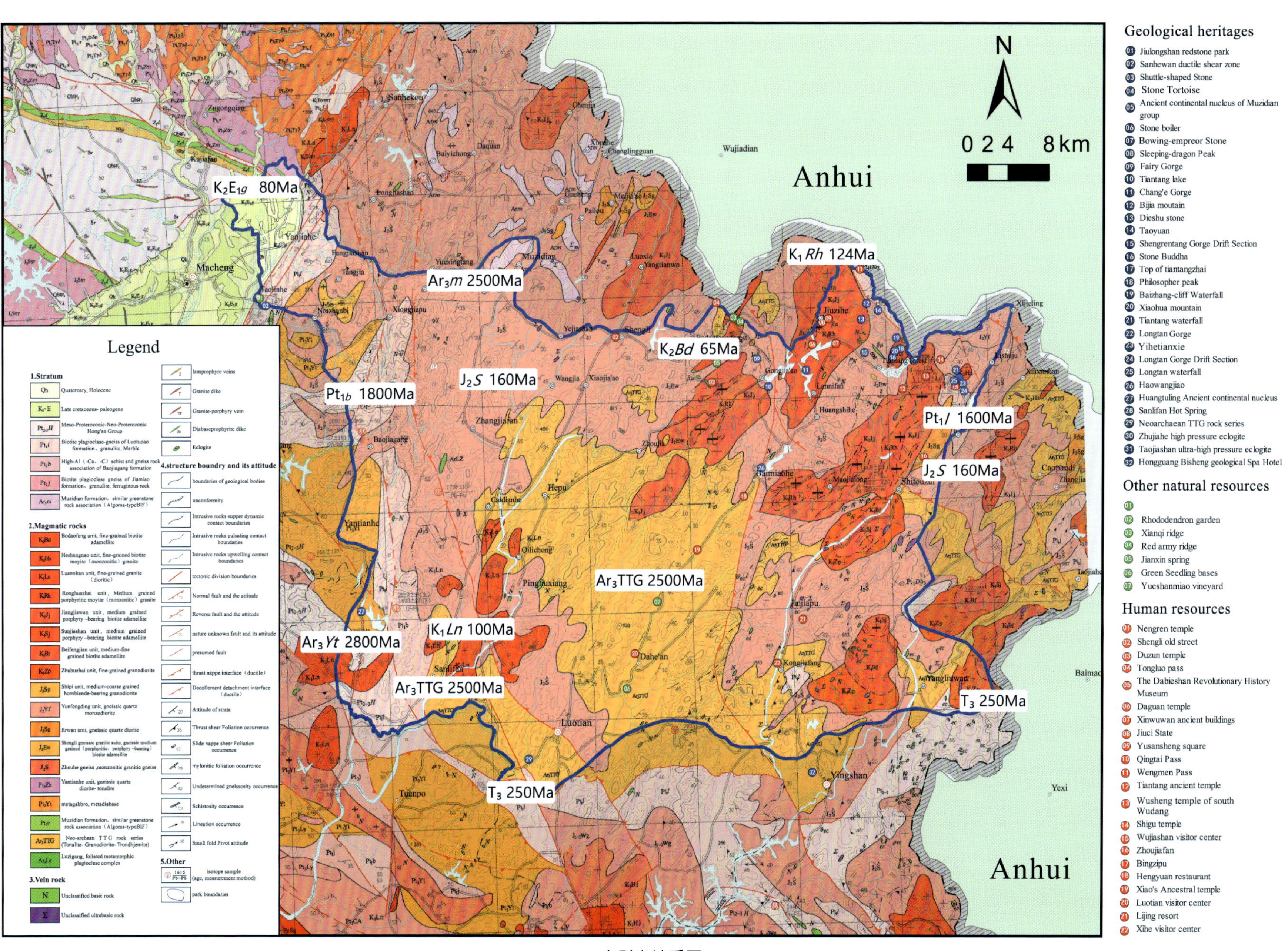

大别山地质图
Geological map of Dabieshan

九龙汇聚 Jiulong（nine dragons）gathering
摄影：林涛 Photographed by Lin Tao

2019年5月11日黄冈大别山世界地质公园揭碑开园

The opening ceremony of Huanggang Dabieshan UNESCO Global Geopark On May 11, 2019

2018年9月13日，被授予黄冈大别山世界地质公园证书
Huanggang Dabieshan UNESCO Global Geopark Certificate awarded on September 13, 2018

2017年7月18日，申报世界地质公园汇报会
The report meeting for Huanggang Dabieshan
UNESCO Global Geopark On July 18, 2017

GLOBAL GEOPARKS NETWORK

CERTIFICATE

The Executive Board of the Global Geoparks Network has approved

Huanggang Dabieshan - CHINA

as

Global Geoparks Network
Institutional Member
For the period 2018-2021

For the Global Geoparks Network Executive Board:
Date 14th September 2018, at Adamello Brenta UNESCO Global Geopark, Italy

GGN President Prof. Nickolas Zouros
Vice-President Ibrahim Komoo
Vice-President Xiaochi Jin
GGN General Secretary Guy Martini
GGN Treasurer Kristin Rangnes

Geoparks

黄冈大别山世界地质公园成员证书
Certificate to Huanggang Dabieshan as Global Geoparks Network Institutional Member

UNESCO GLOBAL GEOPARKS

On the recommendation of the UNESCO Global Geoparks Council, the Executive Board of UNESCO has designated

Huanggang Dabieshan

as a
UNESCO Global Geopark

UNESCO Global Geoparks explore, develop and celebrate the links between their geological heritage and all other aspects of their natural and cultural heritage. They reconnect human society with the history of our planet across 4,600 million years that have shaped every place on Earth and all life that has ever lived here.

黄冈大别山联合国教科文组织世界地质公园荣誉证书
Certificate to Huanggang Dabieshan as a UNESCO Global Geopark

地质公园获奖荣誉墙
Honor wall of the Geopark

"世界地质公园地质遗产保护与可持续发展"高峰论坛
Summit forum on geoheritage protection and sustainable development within the Global Geopark

2019年与墨西哥国家联邦政府环保部签署合作意向书
In 2019, it signed the letter of intent to cooperate with the Ministry of Environmental Protection of the National Federal Government of Mexico

2018年联合国教科文组织世界地质公园年会
The 2018 annual conference of China UNESCO Global Geopark

联合国专家研讨地质公园建设
Evaluators discuss the development of the Geopark
摄影：罗焱超 Photographed by Luo Yanchao

专家在地质博物馆
Evaluators visiting the geopark museum

专家考察地质遗迹点
Evaluators visiting the geosite

德国梅塞尔坑交流　Exchange activities in Messel Pit, Germany

意大利交流2 Exchange activities in Italy No. 2

意大利交流1 Exchange activities in Italy No. 1

黄冈
有遺篇
Huanggang:
Park of Geological Heritage

中央山系——自然遗产 / Central Moutains—Natural Heritage /

黄冈自然人文交相辉映。大别山巍峨磅礴、巧夺天工，连绵境内数百里，其主峰天堂寨海拔1729.13m，集奇、险、幽于一体，堪与泰山、庐山媲美。境内倒、举、巴、希、蕲、华阳河六水并流，百湖千库星罗棋布。

依山傍水、风光秀丽的黄冈吸引了不少名人名士慕名前来，李白、杜牧、王禹偁等历代骚人在此吟咏千古名篇，苏轼因此成就其文学巅峰。

Huanggang enjoys beautiful environment, with natural and cultural landscapes enhancing each other's beauty. Dabieshan Mountain is not only lofty and grand, but also of great momentum. It stretches for hundreds of miles and the main peak, Tiantangzhai, is 1 729.13m above sea level. It is extraordinary, precipitous and peaceful and as beautiful as Mount Taishan and Mount Lushan. Within the territory, there are six rivers flowing named Dao, Ju, Ba, Xi, Qi, and Huayang. And there are hundreds of lakes and thousands of reservoirs.

Huanggang is a beautiful city, which is surrounded by mountains and rivers, attracting many celebrities and famous people to admire its beauty. In history, famous poets such as Li Bai, Du Mu, Wang Yucheng etc. all chanted and sang their outstanding poems here. For example, Su Shi, one of the most famous poets, was at the peak of his literary achivements during his stay in Huanggang City.

雄奇大别山罗田
Magnificent Dabieshan located in Luotian County
摄影：徐原超
Photographed by Xu Yuanchao

花海云潮 Sea of flowers and cloud tide
摄影：胡正平 Photographed by Hu Zhengping

龟峰山
Guifengshan Mountain

薄刀峰奇松1 Wondrous pines in Bodaofeng Peak No.1

薄刀峰奇松2 Wondrous pines in Bodaofeng Peak No.2

大白鹭 Great egret
摄影：周卫烈 Photographed by Zhou Weilie

春到大别山
Spring in Dabieshan

杜鹃花海近景
Close-up view of the sea of azaleas

大别山天堂湖湿地公园博物馆
Museum of Dabieshan Tiantang Lake National Wetland Park

厚重历史—文化遗产 / Profound History—Cultural Heritage /

黄冈地处“吴头楚尾”，是鄂东文化的发祥地之一。吴、楚、汉等古老文化在这里相互交汇融合。其中黄冈市区的东坡赤壁，是全国重点文物保护单位。东坡赤壁的楼阁始建于西晋初年，距今约1700余年，后多次重建，现有面积400余亩。建筑物计有“二堂、三楼、二阁、一斋、一像、一峰、九亭”。其中二赋堂内有一块大木壁，正反面刻着前、后《赤壁赋》全文；留仙阁有一幅苏东坡游赤壁全图；碑阁内有百余块石碑，刻满苏东坡的书法。

Huanggang is located at the "the head of Chu State and the end of Wu State", which is one of the birthplaces of eastern Hubei culture. Ancient cultures, such as Wu, Chu and Han mingle and merge here. Dongpo Red Cliff, situated in Huangzhou Distrct, Huanggang City, was listed as Chinese National Key Cultural Relics Protection Unit. The Pavilion in Dongpo Red Cliff was built in the early Western Jin Dynasty, dating back to about 1700 years ago. After being rebulit for several times, so far the Pavilion covers an area of more than 400 acres. There are two halls, three towers, two chambers, one studio, one stastue, one peak and nine pavillions in Dongpo Red Cliff. There is a large wooden wall in the Double Fu Hall, on the front and back side of which there engraved with the full text of Song of Red Cliff. There is a full figure named The Tour of Su Dongpo in the Red Cliff hung in Liuxian Chamber; And there are hundreds of stone tablets in the stelas pavilion, which are engraved with the calligraphy of Su Dongpo.

东坡赤壁 Dongpo Red Cliff
摄影：方华国 Photographed by Fang Huaguo

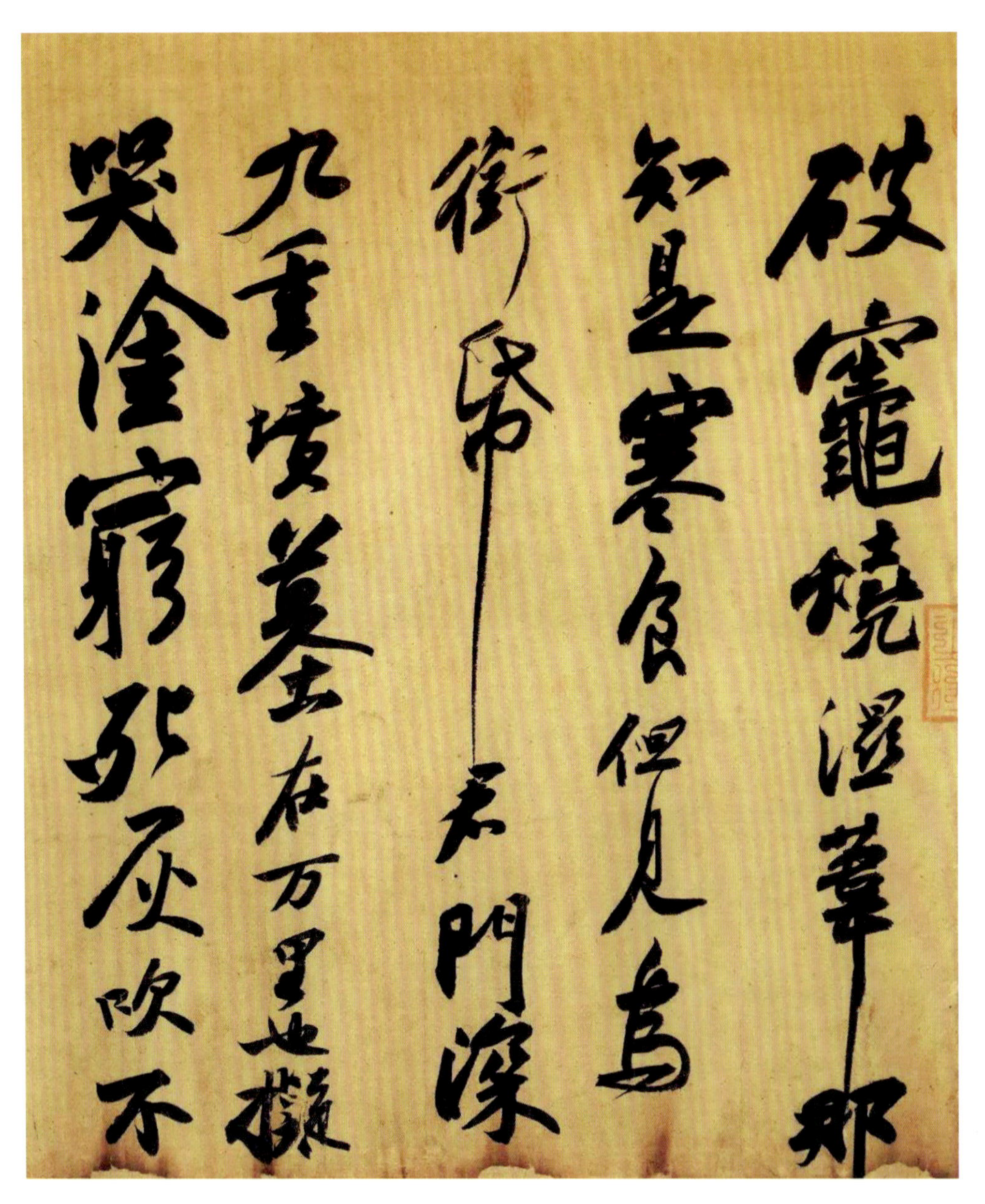

苏轼寒食帖
Sushi's calligraphy—Hanshitie
图片源自：https://baijiahao.baibu.com/s?id=1682694158021066226&wfr=spiolke&for=pc.

东坡文化——寒食林
Dongpo culture—Hanshilin Stele

花岗岩石雕东坡造像
Portrait of Su Dongpo, carved from granite

黄冈遗爱湖国家湿地公园位于湖北省黄冈市区，总面积463.9hm²，其中水域面积378.22hm²，以市区内的东湖、西湖和菱角湖为主体，是长江中下游沿江平原湖区典型湖泊湿地。湿地公园划分为保育区、合理利用区、恢复重建区、宣教展示区和管理服务区五个功能区。2014年3月，遗爱湖被评为“十大湖北最美湖泊”之一；2016年8月16日，通过国家林业局2016年试点国家湿地公园验收，正式成为“国家湿地公园”。

Huanggang Yiai Lake National Wetland Park is located in Huanggang City, Hubei Province, with a total area of 463.9 hectares, including 378.22 hectares of water area. It is mainly composed of East Lake, West Lake and Lingjiao Lake. It is a typical lake wetland in the middle and lower reaches of the Yangtze River. The wetland park is divided into five functional areas: conservation area, rational utilization area, restoration and reconstruction area, mission and teaching exhibition area, management and service area. In March 2014, Yiai Lake was rated as one of the "Ten Most Beautiful Lakes in Hubei province". On August 16, 2016, the wetland park was officially named as Yiai Lake National Wetland Park after being accepted by the National Forestry Administration as an official National Wetland Park.

遗爱湖芸香阁
Yunxiang Chamber in Yiai Lake

黄冈遗爱湖国家湿地公园
Yiai Lake National Wetland Park

遗爱湖梅园1
Plum Garden in Yiai Lake National Wetland Park NO. 1

遗爱湖梅园2
Plum Garden in Yiai Lake National Wetland Park NO.2

根巅在罗田

Luotian: Root and Top of Mount Dabieshan

根巅在罗田 / Luotian: Root and Top of Mount Dabieshan /

黄冈罗田境内地质遗迹数量多、价值高，有年龄超过28亿年的古老变质岩和大片的古老岩浆岩“TTG”岩系，这些被称为“大别山之根”的古老岩石是大别山古陆核最古老的物质，也是地球演化中古陆壳形成的证据，赵鹏大院士曾题词赞誉：“千里大别山，根巅在罗田。”

In Luotian County, there are lots of valuable geosites in this county. For example, there preserved old metamorphic rocks formed more than 2.8 billion years ago and a wide range of ancient magmatic rocks of Trondhjemite-Tonalite-Granodiorite, which are the oldest rocks of Dabieshan continental nucleus, known as "the root of Dabieshan". It is also the evidence for the formation of the ancient continental crust in the evolution of the earth. Zhao Pengda, Academician of Chinese Academy of Sciences, once praised by writing an inscription: “Thousands of miles in Dabieshan, and root It’ s and top in Luotian.”

赵鹏大院士在黄冈大别山世界地质公园考察

Zhao Pengda, Academician of Chinese Academy of Sciences, inspecting Huanggang Dabieshan UNESCO Global Geopark

赵鹏大院士题词
Inscription written by Zhao Pengda, Academician of Chinese Academy of Sciences

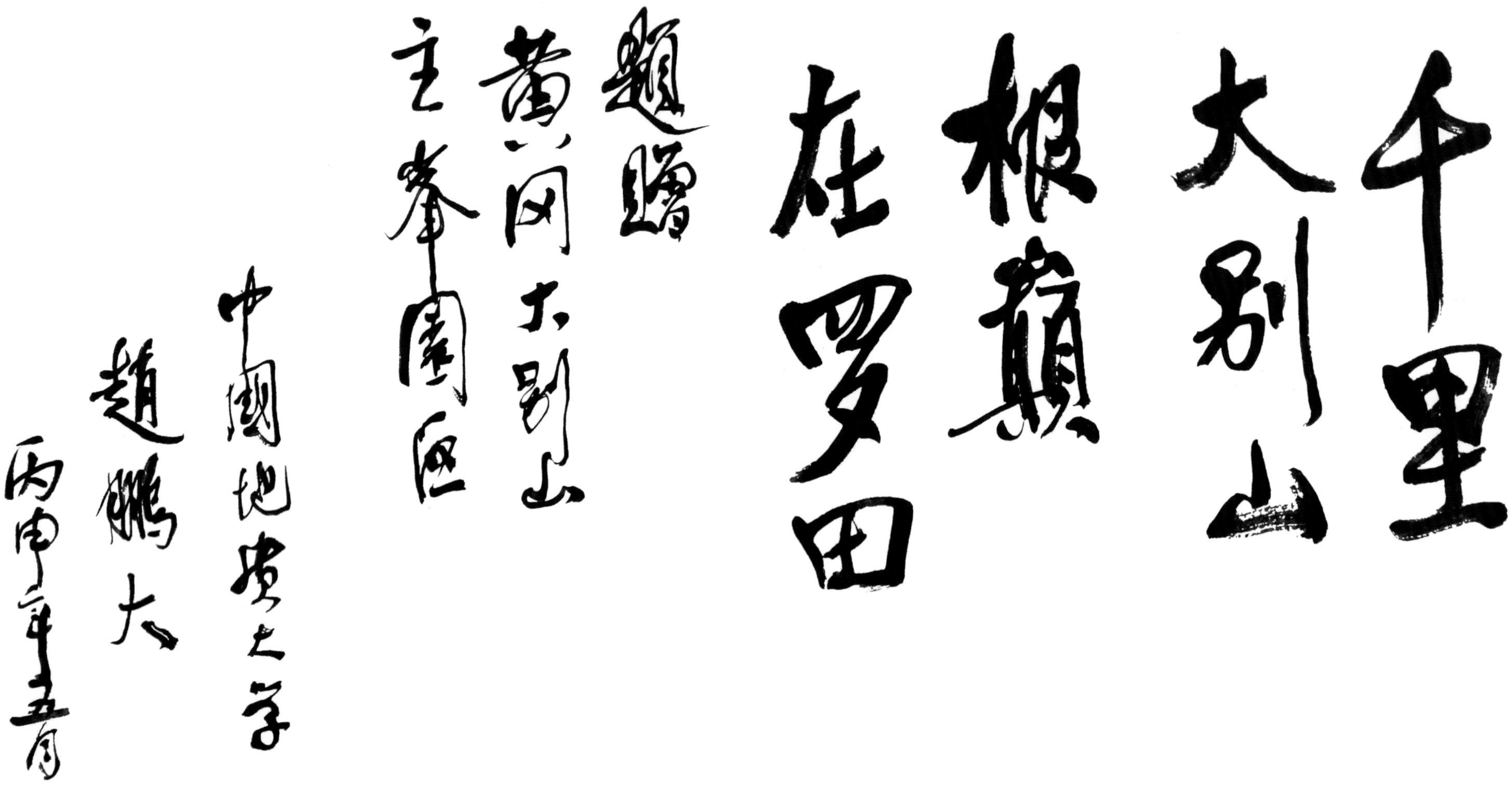

千里大别山
根巅在罗田

题赠
黄冈大别山
主峰园区

中国地质大学
赵鹏大
丙申年五月

赵鹏大院士题词
Inscription written by Zhao Pengda, Academician of Chinese Academy of Sciences

黄土岭古陆核 / Huangtuling Ancient Continental Nucleus /

黄土岭古陆核岩性为紫苏石榴黑云片麻岩，出露仅约10m²，呈不规则团块状，长一般50～70cm，主要矿物为石榴子石、黑云母、紫苏辉石、钾长石、斜长石、堇青石和石英等，具片麻理构造。锆石同位素年龄超过28亿年，是大别山地区最古老的古陆核组成物质。

Huangtuling ancient continental nucleus, with an exposure area of about 10m², is mainly composed of perilla garnet biotite gneissic rock. It is in irregular mass with a length of 50~70cm. The main minerals are garnet, biotite, hypersthene, potash feldspar, plagioclase, cordierite and quartz etc., so it has a gneissic structure. Evidence from zircon isotopic dating suggests that the nucleus has an age of more than 2.8 billion years, and it is the oldest rocks that has formed the Dabieshan continental nucleus.

黄土岭古陆核
Huangtuling ancient continental nuclues

黄土岭紫苏黑云角闪斜长片麻岩显微照
Photomicrographs of perilla biotite hypersthene plagioclase granulites in Huangtuling

黄土岭紫苏石榴黑云片麻岩
Perilla biotite garnet gneiss in Huangtuling

朱家河榴辉岩 / Zhujiahe Eclogite /

朱家河高压榴辉岩成透镜状
Zhujiahe high-pressure Eclogite, shaped in lens

朱家河高压榴辉岩位于罗田城南西朱家河。大别造山带是横亘于华北板块和杨子板块之间的陆、陆碰撞型造山带,这里有全球规模最大，剥露最好、保存最完整的高压—超高压变质岩带。榴辉岩是典型的高压—超高压变质岩，通过岩石学研究，这类岩石对追溯高压—超高压变质作用的历史，了解变质带的形成、演化，以及研究大陆动力学等科学问题，具有重要的意义。

Zhujiahe high pressure eclogite is located in Zhujiahe Village, southwest of Luotian County. Dabie orogenic belt is a continent-continental collision orogenic belt between the north China plate and the Yangtze plate. There preserved high-pressure and ultra-high-pressure metamorphic rock belt which are large in scale, best in denudation and the most complete in preservation in the world. Eclogite is a typical high-pressure and ultra-high-pressure metamorphic rock. This kind of rock is of great significance to trace the history of high-pressure and ultra-high-pressure metamorphism, to understand the formation and evolution of metamorphic belts, and to study continental dynamics and other scientific problems via petrographic study.

罗田朱家河榴辉岩地质剖面
Geological bisection of eclogite in Zhujiahe Village, Luotian County

朱家河高压榴辉岩显微照片
Photomicrograph of Zhujiahe high-pressure eclogite

朱家河高压榴辉岩显微照片
Photomicrograph of Zhujiahe high-pressure eclogite

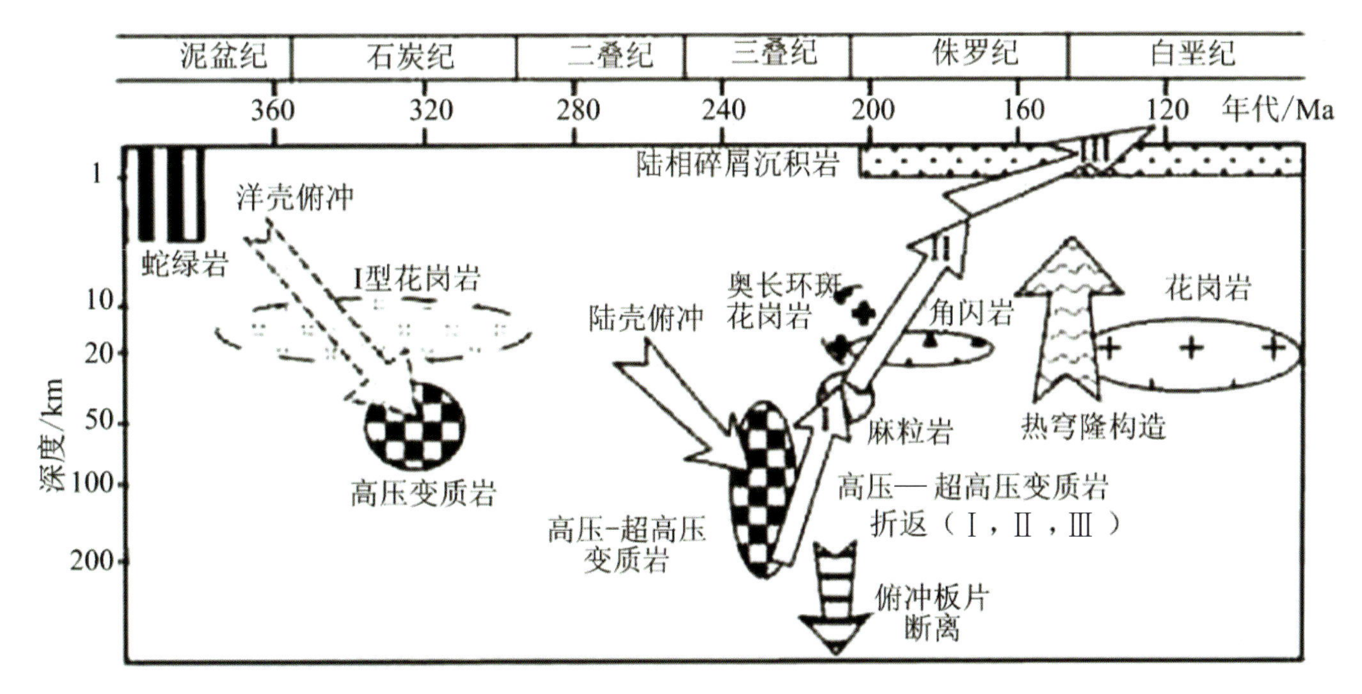

榴辉岩形成机理示意图 Schematic diagram to the formation mechanism of eclogite
(据王清晨, 2020 According to Wang Qingchen, 2020)

不忘初心 牢记使命
ETC
收费站
货车靠右
超限超载检测
49t
5m
禁止车速低于
60...的车辆入内
20

石趣园全景 Panorama of Stone Garden
摄影：华仁 Photographed by Hua Ren

石趣园奇石1

Rare stones in Stone Garden No.1

石趣园奇石2
Rare stones in Stone Garden No.2

岩石细节
Rock details

TTG岩系 / TTG Rock Series /

地质解说牌 Geological interpretation panels

黄冈大别山世界地质公园及周边区域，出露国内罕见的距今28亿年的古老变质岩——紫苏石榴黑云母片麻岩和大片的原始造陆花岗侵入岩——TTG岩系，即：英云闪长岩、奥长花岗岩、花岗闪长岩三类岩石的组合。这些被称为“大别山之根”的古老岩石是大别山古陆核最古老的物质，也是地球演化中古陆壳形成的证据。

In Huanggang Dabieshan UNESCO Global Geopark and its surrounding areas, there exposed the rare ancient metamorphic rocks dating back 2.8 billion years in China, perilla garnet biotite gneissic rock and a large scale of primitive continental granitic intrusive rocks which includes Trondhjemite, Tonalite and Granodiorite. These ancient rocks, known as "the root of Dabieshan", is also the evidence for the formation of the ancient continental crust in the evolution of the earth.

TTG岩系罗田栗子坳河地质遗迹全景 Panoramic View of TTG rock series, Lizi'ao River in Luotian County
摄影：华仁 Photographed by Hua Ren

罗田TTG脉体和基体
Vein and matrix of trondhjemite-tonalite-granodiorite in Luotian County

罗田TTG岩体肠状褶皱
Ptygma of Trondhjemite-Tonalite-Granodiorite in Luotian County

罗田TTG岩体“旋转眼球”
“Rotating eyeball” stucture of Trondhjemite-Tonalite-Granodiorite in Luotian County

TTG岩系科普宣传 Interpretation panels of TTG rock series
摄影：华仁 Photographed by Hua Ren

天堂湖国家湿地公园

/ Tiantang Lake National Wetland Park /

中华秋沙鸭 Chinese merganser
摄影：周卫烈 Photographed by Zhou Weilie

天堂湖湿地 Tiantang Lake Wetland
摄影：王珂 Photographed by Wang Ke

天堂湖湿地夕阳 Sunset of Tiantang Lake wetland
摄影：王珂 Photographed by Wang Ke

大别脊梁薄刀峰

/ Bodaofeng Peak—The Ridge of Dabieshan /

薄刀峰位于罗田县东北部,在大别山主峰南麓山脉的一条山脊之上，其两侧为悬崖峭壁，山脊形如刀刃，故名“薄刀峰”。薄刀峰以花岗岩地貌为主，造景岩石形成于晚白垩世(6500万年前)，长10余千米，北东向高山峻岭，蜿蜒于鄂皖边境。在大别山造山带核部的隆起过程中，历经岩浆侵入、构造抬升、断块活动、风化剥蚀等一系列地质演化，形成了高、陡、险、峻的“薄刀”状山脊花岗岩地貌，犹如天地间的兵器锻造带。

Bodaofeng Peak is located in the northeast of Luotian County, on a ridge at the southern foot of the main peak of Dabieshan. It is flanked by steep cliffs and the ridge looks like a blade, hence the name "Bodao Peak". Bodao Peak is mainly composed of granite, formed in the late Cretaceous (65 million years ago). This peak is more than 10 kilometers long, the northeast part of which is steep and winds to the border of Hubei and Anhui province. The granite landscapes of Bodao Peak is caused by the intrusion of magma and then a series of geological evolution, such as tectonic uplift, fault-block movement, weathering denudation etc.,happened during the uplifting process of Dabie orogenic belt, after which there formed a granite ridge, like a thin knife. It is high, steep, dangerous and beatutiful, like a weapon belt forged between the heaven and the earth.

薄刀峰锡锅顶
Xiguoding in Bodaofeng Peak

薄刀峰天子弯腰
Tianzi Wangyao in Bodaofeng Peak

生态乐园薄刀峰 Ecological Paradise Bodaofeng Peak
摄影：周红林 Photographed by Zhou Honglin

天堂寨 / Tiantangzhai /

在扬子板块与华北板块相互碰撞汇聚造山过程中，地壳隆升，岩石变形，导致大规模隆起，形成山系，其后又经过常年风化、剥蚀、地表水的侵蚀和重力崩塌等多种内外地质作用的叠加，形成中国中央山系，也造就了如今的天堂寨。天堂寨群峰逶迤绵亘，山势雄伟险峻，为典型的花岗岩山岳地貌景观，以“山雄、壑幽、水秀”著称，占地面积120km^2，是黄冈大别山世界地质公园的核心景区之一，也是国家级森林公园和自然保护区。大别山顶峰天堂寨海拔1 729.13m，号称“中原第一峰”，是长江和淮河的分水岭，登上顶峰可一眼观安徽、湖北两省以及罗田、英山、金寨三县。

During the process when the Yangtze plate and north China plate collided with each other and then the earth's crust was uplifted, the rocks were shaped, and hills swell on a grand scale, thus formed the mountain Dabieshan. After years of weathering, denudation, surface water erosion and gravitational collapse, accompanied by inner and outer geological dynamics, there formed the central mountain range of China. This is also the formation of Tiantangzhai. The mountain peaks of Tiantangzhai are mainly composed of granite and stretches for miles, which is majestic and steep, famous for its "magnificent mountains, secluded valleys and clear water". Covering an area of 120km^2, it is one of the core scenic area of Huanggang Dabieshan UNESCO Global Geopark, which is also part of national forest park and natural reserve. Tiantangzhai, the summit of Dabieshan, known as "The First Peak of Central China", is 1 729.13m. Tiantangzhai is the watershed between the Yangtze River and the Huai River. While climbing on the top of Tiantangzhai, the tourists can enjoy the view covering the beautiful scenery of Luotian, Yingshan and Jinzhai counties within the territory of Anhui and Hubei province at a glance.

天堂寨一脚踏两省三县

A foot treading on three counties within two provinces in Tiantangzhai

摄影：舒胜前

Photographed by Shu Shengqian

古杜鹃树
Ancient rhododendron tree

天堂寨主峰
The main peak of Tiantangzhai

笔架山 Bijiashan Mountain

鸠兹古镇 Jiuzi Town

千里大别山，美景在罗田
Thousand miles of Dabieshan, fantastic view in Loutian

家学展馆
Pavilion of Family Learning

罗田甜柿第一乡
Luotian ,the first sweet persimmon town

大别川画廊之秋 Autumn of Dabiechuan Gallery
摄影：华仁 Photographed by Hua Ren

Yingshan: A panoramic View of Mountains

大别山主峰

/ The Main Peak of Dabieshan /

号称“中原第一峰”的大别山主峰天堂顶，海拔1 729.13m，被誉为华中地区的“绿色明珠”。天堂顶的地质遗迹有距今6500万年和距今1.24亿年的花岗岩。这些位于地壳内部的岩石经构造运动向上提升，隆起成山，暴露出地表后，又经过风化、雨水冲蚀、重力崩塌等地质作用，最终形成现今的花岗岩地貌。登上大别山主峰，巍巍群山尽收眼底，气势磅礴，这里是登高、观赏日出和欣赏云海的绝佳之地。

The main peak of Dabieshan, named Tiantangding, is known as "The First Peak of Central China", with an altitude of 1 729.13m. It is also honored as the "green pearl" in Central China. The geosites of Tiantangding are granite, which were formed between 65 million years ago and 124 million years ago. At the first beginning, these rocks were located in the inner crust of the earth, they were uplifted by tectonic movement, and then formed the mountain. When exposed to the surface, they have undergone a series of geological process, such as weathering, rain erosion, and gravitational collapse, eventually there formed the current granite landform. Climbing to the main peak of Dabieshan, you will have a panoramic view of the towering mountains, which is majestic and magnificent, being an excellent sight for watching sunrise and clouds.

顶峰栈道
Plank Road at the top of the mountain
摄影：舒胜全
Photographed by Shu Shengquan

英山玻璃栈道
Glass walkway in Yingshan County

冬晨天堂寨
The morning of winter in Tiantagnzhai

主峰全貌
Panorama of the Main Peak

云雾缭绕 Clouds and mist
摄影：刘志雄 Photographed by Liu Zhixiong

南武当耍剑 Swordplay in South Wudang
摄影：陈国华 Photographed by Chen Guohua

南武当武圣宫 Wusheng Temple of South Wudang
摄影：张新安 Photographed by Zhang Xin'an

南武当全景
Panorama of South Wudang

龙潭河谷

/ Longtan Gorge /

龙潭河谷位于大别山主峰园内，平均海拔390m，面积50m²，河谷深切，水流丰沛，生态原始，地质遗迹多样。龙潭河谷拥有九潭十八瀑，峡谷幽深险秀，崖险壁峻，怪石嶙峋，被誉为“华中河谷第一景”；因峡谷穿越十多亿年的岩浆活动，河床上纵横交错的脉体七彩纷呈，又被称为“七彩溪”。龙潭河谷内著名的景点有“银河天泻”“乌龙戏水”“龙潭飞瀑”等。

Longtan Gorge, with an average elevation of 390 meters and an area of 50 square meters is located in the main peak of Dabieshan. The steep valley is filled with water and covered with natural environment, and there preserved rich geological relics. Being deep and quiet, precipitous but beautiful, the Valley has nine pools and eighteen waterfalls, which are full of steep cliffs and rugged rocks, known as the "The First River Valley in Central China"; The valley has undergone quantity of magmatic activity, which has been lasted for billions of years, so the river bed is crisscrossed with colorful veins, hence the name "Colorful Streams". There are many famous scenic spots in Longtan Valley, such as "Milky Way" "Long Playing in Water" "Longtan Waterfall" and so on.

龙潭峡谷
Longtan Gorge

河谷激流 Water coming down in torrents
摄影：彭世海 Photographed by Peng Shihai

龙潭河谷碎石1
Gravels in Longtan Gorge No. 1

龙潭沟 Longtan ditch

龙潭河谷瀑布
Waterfall in Longtan Gorge

龙潭河谷碎石2 Gravels in Longtan Gorge No. 2

龙潭河谷变质岩

/ Longtan Gorge Metamorphic Rocks /

龙潭河谷地质遗迹种类丰富，长约10km的峡谷，其岩石穿越了16亿—1亿多年地质时空，峡谷内可见到古元古代、中侏罗世、晚白垩世的岩石各种各样的岩石。龙潭沟河谷及河床出露有16亿年前的古元古代大别山群黑云斜长片麻岩，峡谷的上部岩体属1.7亿年前的中侏罗世片麻状二长花岗岩和片麻状石英闪长岩，除此，还出露有6500万年前晚白垩世的基性侵入岩脉。

The Longtan Gorge is about 10 kilometers in length and has a rich variety of geological relics.The rocks were formed in several geological ages, from 1.6 to 100 million years ago, so multiple types of rocks can be seen in the gorge, for example, rocks formed during the Paleoproterozoic, Middle Jurassic and Late Cretaceous period. The exposed rocks in Longtan Gorge and its riverbed belongs to Dabieshan group biotite plagioclase gneiss, which were formed during the Paleoproterozoic period dating back to 1.6 billion years ago, and the upper rock mass of the gorge are gneissic adamellite and quartz diorite, which were formed 170 million years ago during the old Middle Jurassic, In addition, there also exposed basic intrusive dike formed 65 million years ago during the Late Cretaceous period.

英山龙潭河谷黑云斜长片麻岩及后期互穿岩脉

Biotite plagioclase gneiss and the vein interpenetrating with each other in Yingshan Longtan Gorge

龙潭棒槌石
Bangchui stone in Longtan Valley

龙潭沟16亿年前的黑云斜长片麻岩
Biotite plagioclase gneiss in Longtan ditch, formed 1.6 billion years ago

黑云斜长片麻岩及后期剪切带
Biotite plagioclase gneiss and later shear zone

黑云斜长片麻岩韧性剪切带及眼球构造
Ductile shear zone and eyeball structure of biotite plagioclase gneiss

黑云斜长片麻岩韧性剪切带小褶皱
Small folds in ductile shear zone of biotite plagioclase gneiss

黑云斜长片麻岩韧性剪切带及牵引构造
Ductile shear zone and traction structure of biotite plagioclase gneiss

西河十八湾

/ 18 Bends of Xihe River /

“西河十八湾，湾湾不一般。”西河，是大别山腹地英山县西部的一条河流，是浠河主要支流，发源于安徽省金寨县长岭乡曾家山西界岭，由北向南流经四镇两乡（长岭乡、石头咀镇、金家铺镇、孔家坊乡、红山镇、温泉镇），在县城两河口与东河汇合，在九垄口村流入白莲河，全长59km。她像一束银色的丝带在阳光下灿然闪烁，在两岸郁郁葱葱的树木之间，飞星溅沫，逶迤在大别山之间，环环弯弯，绘成了美丽西河十八湾。

"There are several bends in Xihe River and each is different from one to another". Xihe is a river which is located in the hinterland of Dabieshan and lies in the west of Yingshan County. It is a major tributary of the Xi River orginating from Xijie Ridge, Zengjia Mountain, Changling Village, Jinzhai County, Anhui Province and flows through four towns and two villages from north to south(Changling Village, Shitouzui Town, Jinjiapu Town, Kongjiafang Village, Hongshan Town and Wenquan Town), and merges with Donghe in Lianghekou Village of Yingshan County, and then flows into Bailianhe at Jiulongkou Village, which measures a total length of 59 kilometers. Shining brightly in the sun, the whole river looks like a bunch of silver ribbons. With blooming trees in both sides, the river is winding along Dabieshan, encircling and bending, thus painted beautiful villages embraced by Xihe River.

美丽西河 Beautiful Xihe River

张咀水库
Zhangzui Reservoir

英山温泉 / Hot Spring in Yingshan County /

英山县是著名的温泉之乡，《英山县志》(1998年12月)将“温泉春景”作为英山八景之一。英山县城依山傍河，有东南西北4处温泉，其温泉分布之广、流量之大、水质之优，十分罕见。经鉴定，英山县内的温泉流量大、水质软、碱度低，在温泉中含有多种元素，有着较高的开发价值，完全可用于工业、农业、医疗、科研、体育以及为人们生活服务，应用前景十分广泛。

Yingshan County is well-known for its hot springs. In *The Old Yingshan County Chronicle*, hot springs in the spring is recorded as one of The Eight Scenic Spots in Yingshan County, surrounded by mountains and rivers. There are 4 hot springs in all directions. In Yingshan County, hot springs are widespread, and known for its large flow and clear water, which is exceedingly rarely to see. It has been identified that the flow of hot springs in Yingshan County is large, the water has been softening with low alkalinity, besides there contains multiple elements, thus it is of high development value, which can be used in industry, agriculture, health care, scientific research, sports, and even be served for people's lives, so the application prospect is very broad.

四季花海花乐汤温泉
Letang Hot Spring in Four Seasons of Flower Sea

洪广温泉酒店
Hongguang Hot Spring Hotel

洪广温泉 Hongguang Hot Spring
摄影：章卫军 Photographed by Zhang Weijun

麻城 赏杜鹃

Maocheng: Azalea Bloom in Spring

最美地质路——桃林河（S309-木子店段）

/ The Most Beautiful Geological Highway—Taolinhe (S309 Macheng Muzidian) /

麻城最美地质公路起始于公园西边的麻城闫家河桃林村。行走在“最美的地质公路上”，沿途可考察构造运动和物理风化、崩塌和地应力作用在这里留下的诸多印迹。在桃林河，可游览地质文化村；在三合湾糜棱岩带小游园，探究迄今为止在中国东部发现的最大的韧性剪切带，也是大别山造山过程中动力及运动的轨迹；走过向明红叶村、十人寨，感受旅游休闲村庄的温度；寻访木子店，揭秘大别山古陆核，识28亿年前大别山遗留的绿岩带的物质组成；登上龟峰山，仰望“天下第一龟”的神奇，徜徉于万亩杜鹃花海。

The most beautiful geological highway starts from the Macheng Yanjiahe Taolin Village, west of the park. Walking along the "The most beautiful geological highway", we can observe many marks left by tectonic movement and physical weathering, collapse and ground stress. In Taolinhe Village, we can visit the geological culture village; In Sanhewan Mylonite Zone Small Garden, we can explore the largest ductile shear zone found so far in eastern China ,which is also the dynamic and motion track of Dabieshan orogeny. Walking through Xiangming Hongye Village, Ten-persom village, we can feel the warm of tourism leisure farm.To visit Muzidian and find out the ancient continental core of Dabieshan and the material composition of the greenstone belt left by Dabieshan from 2.8 billion years ago. To climb Guifengshan, admire the magic of "The Number One Stone Tortoise", wander in the sea of rhododendron over 10000 mu. fragrant and sweet,,and to experience a free and desirable life once led by the peach blossom fairy.

最美公路变形带1
Geological Deformationbelt by the Most Beautiful Highway No.1

最美公路变形带2
Geological Deformationbelt by the Most Beautiful Highway No.2

鹭舞桃林河 Egret dancing in Taolinhe
摄影：胡正平 Photographed by Hu Zhengping

熊家铺风光

/ Scenery of Xiongjiapu Town /

在这座古朴的小镇中，能够欣赏到古色古香的房屋建筑，逛一逛琳琅满目的商品铺子，品一杯芳香四溢的茗茶，重温昔日民间商贸一条街的繁荣景象。

Xiongjiapu is a quaint town, in which the tourists will have a chance to admire quaint buildings and to stroll assortment of shops, or to drink a cup of tea, tasting its rich aroma, as if the tourists were in the booming scene of the business street in former days.

熊家铺小镇1 Scenery of Xiongjiapu Town No.1

熊家铺风光 Xiongjiapu scenery

摄影：胡正平 Photographed by Hu Zhengping

熊家铺小镇2 Scenery of Xiongjiapu Town No.2

熊家铺小镇3 Scenery of Xiongjiapu Town No.3

秋染熊家铺 Autumn of Xiongjiapu Town
摄影：胡正平 Photographed by Hu Zhengping

龟峰山 / Guifeng Mountain /

龟峰山园区位于大别山世界地质公园的西部，距麻城市区约25km。主峰高约1300余米，因其地形山势酷似一只翘首苍天的巨形神龟，堪称世界地质奇迹，被誉为“天下第一龟”，故得名“龟山”，又称“长寿山”，是大别山中的名山，也是神秘宇宙恩赐给人类的自然瑰宝。龟峰山山顶怪石险峻，雄伟绮丽。龟峰旭日、仙人脚、望儿石、天梭石、试剑石等象形山石具有重要的科学研究价值。

Guifengshan area is located in the west of Huanggang Dabieshan UNESCO Global Geopark, with a distance of 25 kilometers away from Macheng City. The main peak is about 1300 meters above sea level. It is named as Guifengshan because of its terrain, shaped like a giant turtle heading up to the sky, which can be called as a geological miracle in the world and it is honoured as the greatest turtle in the world, hence the name Guifengshan (Tortoise peak), also know as Longevity Mountain(Because Tortoise is a symbol for long life in Chinese). Guifengshan is not only a famous mountain in Dabieshan, but also a nature's gift to mankind in the mysterious universe. The rocks on the peak of Guifengshan is steep and majestic, precipitous and beautiful, and there preserved a number of pictographic rocks, such as Guifeng Rising Sun, Immortal Foot, Wanger Stone, Shuttle-shaped Stone, and Sword-practicing Stone, which are of important scientific research value.

母子情深 Mother and son affection

龟峰山小镇
Guifengshan Town

片麻状花岗岩 The gneissic granite

大别春色 Spring of Dabieshan
摄影：刘志雄 Photographed by Liu Zhixiong

杜鹃花海

/ Sea of Rhododendron /

“踏春赴麻城，山中赏杜鹃。”在海拔700～1300m的山岭上，冬暖夏凉，降水充沛，风化的岩浆岩形成偏酸性的疏松土壤，极利于杜鹃花的生长。这里形成了分布集中、保存完好、群落结构完整、株型优美、总面积达6670万km²的原生态古杜鹃植物群落。每逢暮春，漫山遍野的古杜鹃红艳似火，犹如彩霞绕林，被上海世界吉尼斯总部评为“中国面积最大的古杜鹃群落”。

"If you go for a spring walk in Macheng, you will enjoy a sea of rhododendrons in Guifengshan." On the mountains 700～1300m above sea level, the temperature is warm in winter and cool in summer, which delivers an intense abundance of water; and weathered magmatic rocks can form a kind of soil, acidic and loose, which is extremely conducive to the growth of rhododendrons, hence there formed an original ecological community of ancient rhododendron plants with a concentrated distribution and a good condition. The rhododendron 93 community with a total area of 66.7 million square kilometers, is in complete structure and has a beautiful plant type. In late spring, the tourists would have seen brilliant rhododendron flowers all over the mountainside, as red as fire, like rosy clouds surrounding the forest. And it was named as "The Largest Ancient Rhododendron Community in China" by the Shanghai World Guinness Headquarters.

岭上开遍映山红 Azalea all around the ridge
摄影：王晓霞 Photographed by Wang Xiaoxia

霞映群山 Sunset over Mountains
摄影：邱东旭 Photographed by Qiu Dongxu

龟峰山日出 Sunrise of Guifengshan

摄影：刘志熊 Photographed by Liu Zhixiong

九龙山 / Jiulongshan /

九龙山位于麻城市东北约10km处，为典型的丹霞地貌景观，同时也是黄冈大别山世界地质公园中唯一的沉积岩地貌景观。这里沉积有距今1亿年左右的白垩系典型红层剖面，主要岩性为紫红色砂砾岩，以花岗岩砾石和粗砂岩、泥质胶结为主，岩层以单斜地层产状产出。

九条山岗由中心向四周弯曲蜿蜒，好似九条巨龙盘旋环绕，民间俗称“九龙缠顶”。九龙山脊上建有柏子古塔，已有1300年的历史，塔旁有“唐王洞”“龙井”“寺院”等。明代思想家李贽讲学著述的龙潭寺、钓鱼台遗址，被列为省重点文物保护单位。九龙山主要景点有：白垩系红层剖面及丹霞地貌景观、九龙湖、点将台、救驾桥、烽火台、千佛洞、唐王洞、柏子古塔、九龙寺等。

Jiulongshan is located in the northeast of Macheng City, with a distance of about 10 kilometers. It is a typical Danxia landform and is also the only sedimentary rock landform in Huanggang Dabieshan UNESCO Global Geopark. There deposited typical Cretaceous red bed bisect, formed about 100 million years ago and the main lithology of which is purple-red glutenite, mainly composed of granite conglomerate, gritstone, and shale cementation. The rock formations are produced in the form of monoclinic strata.

Jiulongshan meanders to all around from the center, as if nine giant dragons hovered over towards the blue sky, thus are commonly known as "Nine hovered Dragons". On the ridge of Jiulongshan, there built a tower named Baizi, with a history of 1300 years, around which there built some culture relics, such as "Tangwang Cave" "Dragon Well" "Temple" and so on. Besides there also built some historic relics, named Longtan Temple and Diaoyu platform, which are listed as Provincial Key Cultural Relics Protection Unit, for that Li Zhi, the thinker of the Ming Dynasty once gave lectures and wrote there. The main attractions of Jiulongshan include: Cretaceous red bed bisect and Danxia landform landscape, Jiulong Lake, Dianjiang platform, Jiujia bridge, Beacon Tower, Thousand Buddha Cave, Tangwang Cave, Baizi Tower, Jiulong Temple, etc.

九龙山柏子塔 Baizi Tower in Jiulongshan
摄影：张文科 Photographed by Zhang Wenke

九龙山风光 Scenery of Jiulongshan
摄影：王晓霞 Photographed by Wang Xiaoxia

九龙山砾岩1
Conglomerate in Jiulongshan No.1

九龙山砾岩2
Conglomerate in Jiulongshan No.2

桃林河韧性剪切带及糜棱岩

/ The Ductile Shear Zone and Mylonite in Taolinhe /

始于2亿多年前的印支—燕山运动，使扬子板块和华北板块碰撞，将大别山逐步抬起，直到大约1亿年前后，大别山完成了造山过程。在造山运动中，所发育的北西、北东两个方向高角度走滑型韧性剪切带，规模巨大，分带明显，举世罕见，也是国内迄今为止发现的规模最大的韧性剪切带之一，揭示了大别山造山运动的轨迹和动力。

The Indochina-Yanshan movement, happened more than 200 million years ago caused the collision between the Yangtze plate and the North China plate, after which Dabieshan was gradually lifted . It was not until about 100 million years ago that the Dabieshan has completed its orogenic process. During the orogeny, there developed 2 high-angle ductile shear zones of the strike slip type, which are towards to the northwest and northeast directions respectively. Those 2 ductile shear zones are large in scale and of typical zones, thus it is really rare in the world. It is also one of the largest ductile shear zones discovered so far in China and there reveals the trajectory and dynamics during the Dabieshan orogenic movement.

韧性剪切带1 Ductile shear zone No.1
摄影：彭世海 Photographed by Peng Shihai

韧性剪切带2 Ductile shear zone No.2
摄影：彭世海 Photographed by Peng Shihai

三合湾韧性剪切带及糜棱岩带

/ The NE Trending Ductile Shear Zone and Mylonite zone in Sanhewan /

三合湾韧性剪切带位于桃林河旅游信息咨询中心对面，长40km，宽2～5km，总体走向北偏东20°～30°，南偏西15°～35°。该剪切带原本发育在地壳以下10～15km处，后来在强烈的外力作用下主轴方位发生不可恢复的形变，又经过造山运动抬升、剥蚀等作用露出地表，现今已有2亿多年。

The NE Trending ductile shear zone in Sanhewan village is located on the opposite side of the Taolinhe Tourist Information Center. The zone is 40 kilometers in length and 2～5 kilometers in wide, general strike in 20°～30° north-east and 15°～35° west-south. The shear zone originally developed 10 to 15 kilometers below the crust. Under the strong external force, the main axis was deformed irreversibly, and it was exposed to the surface through orogenic uplifting and denudation, happened 200 million years ago.

三合湾变形褶曲 Sanhewan deformation fold

三合湾粗糜棱岩 Sanhewan adamellitic coarse-grained myliolte

三合湾剪切带褶皱 Sanhewan shear zone fold

桃林河韧性剪切带

Taolinhe ductile shear zone

麻城芦家河剪切褶皱
Shearing fold in Lujiahe, Macheng City

麻城陆家河韧性剪切带
Ductile shear zone in Lujiahe, Macheng City

韧性剪切带
Ductile shear zone

剪切带旋转眼球结构
Rotating eyeball structure of ductile shear zone

超糜棱岩静态重结晶
Static recrystallization of ultramylonite

陆家河糜棱岩变形带
Lujiahe mylonite deformtion band

陆家河糜棱岩变形带与变形鱼
Lujiahe mylonite deformation band and deformed fish

陆家河糜棱岩变形
Lujiahe mylonitic rock deformation

岩浆岩包体 / Magmatic Inclusion /

岩浆岩包体的主要成分为花岗岩，其花岗岩岩体中含斜长角闪岩，为早期片麻岩包裹体。包体形态各异，有棱形和长条形，颜色为黑色。这些包裹体的成长与木子店组绿岩带的物质类似，是岩浆从地下侵入把地球深处形成的基性—超基性岩包裹住，然后带出地表。该类成长过程反映了不同期次的岩浆活动，展现了地球深处的信息，即后期岩浆捕虏早期地球物质信息。

The main component of magmatic inclusion is granite the rock mass of which is plagioclase amphibolite, an early gneiss inclusion. With prismatic and long strips, the inclusions have different shapes and black color. The growth of these inclusions is similar to formation of the Muzidian formation greenstone belt.When the magma intrudes from the ground to depths of the earth, it envelops the basic-ultrabasic rocks and then brings them out of the surface. This kind of growth process reflects magma activities of different periods, and shows information on movement of magma deep in the earth, that is, the information showing the early earth material captured by the later magma.

麻城岩浆岩包裹体1 Macheng magmatic inclusions No.1

麻城岩浆岩包裹体3 Macheng magmatic inclusions No.3

麻城岩浆岩包裹体2 Macheng magmatic inclusions No.2

麻城岩浆岩包裹体4 Macheng magmatic inclusions No.4

木子店古陆核 / Muzidian Ancient Continent Nucleus /

木子店“古陆核”位于麻城市木子店镇东1.5km的公路拐弯处，是出土的可见面积约50m²的残留体。该露头的深色岩石为基性—超基性变质岩石，是距今约28亿年的新太古代岩层的残留体，是大别山最老的石头，也是大别山“古陆核”的重要组成部分，被称为“大别山之根”，是揭开地球早期演化史的钥匙。

The Muzidian ancient continent nucleus is located on a bend in the highway, which has a distance of 1.5 kilometers east of Muzidian Town, Macheng City. The residual body has a visible area of about 50 square meters. The outcrop in dark color is ultrabasic metamorphic rock, which is the remnant of Neoarchean strata formed about 2.8 billion years ago. It is the oldest rock in Dabieshan and an important component of the Dabieshan ancient continent nucleus, known as the root of Dabieshan. It is the key to the early evolution of the earth.

木子店古陆核 Muzidian ancient continental nucleus

麻城木子店古陆核 Muzidian ancient continental nucleus of Macheng City

木子店斜辉橄榄岩显微照片 Micrographs of Muzidian kylite

木子店基性岩包体显微照 Micrographs of Muzidian basic enclave

麻城木子店绿岩带 Muzidian greenstone belt of Macheng City

片麻岩围岩 Gneiss wall rock

龙鳞生辉 Shining dragon scales (Solar panels near Jiulongshan)
摄影：周敦开 Photographed by Zhou Dunkai

原麝 林中跃

Paradise of Musk Deer

密林探索 Forest exploration
摄影：华仁 Photographed by Hua Ren

幽泉
Deep spring

大别山之秋 Autumn of Dabieshan
摄影：汪金元 Photographed by Wang Jinyuan

依山傍水

/ Beside Mountains, Beside Rivers /

黄冈大别山在大地构造上处于华北板块和扬子板块的结合带，地形地貌复杂，气候温和湿润，四季分明，蕴藏着丰富的生物，是中国七大基因库之一，具有很高的科学价值。

Huanggang Dabieshan is located in the joint zone of the North China plate and the Yangtze plate geotectonically. With complex topography, the temperature within Dabieshan is mild and humid climate. Since it has four distinctive seasons, there has bred rich creatures, known as one of the seven gene banks in China, which is of high scientific value.

千年古树 / Thousand-Year Old Tree /

自1亿多年前，大别山在造山运动的作用下，形成了现今的山体基本轮廓，基本保持温暖湿润的气候。因此，在古近纪—第四纪冰期，大别山成为很多植物的避难所，保存了较多的古老孑遗植物和特有种属。这里有国家森林公园和省级森林公园10多处，国家自然保护区1处。森林覆盖率达到90%以上，森林群落保存完好，被誉为华中地区“绿色明珠”。公园内动植物资源十分丰富，其中野生维管植物1461种。例如，有蕨类植物27科49属82种，裸子植物8科16属22种，被子植物160科598属1357种。大别山五针松被列为国家珍稀树种第一批二级保护树种，国家Ⅱ级重点保护野生植物，仅分布于安徽、湖北两省交界海拔700～1200m的中山地带。除了蕨类植物、裸子植物，大别山的被子植物也非常繁盛，如银杏、南方红豆杉、榉树等。

Dabieshan formed the basic outline of the mountain since more than 100 million years ago under the action of orogenic movement. It basically maintained a warm and humid climate, therefore, during the ice period from the Paleogene to the Quaternary, the Dabieshan area has became a refuge for many plants, thus there preserved many ancient relic plants and endemic species. There are more than 10 national forest parks and provincial forest parks and 1 national nature reserve. The forest coverage rate has reached more than 90%, and the forest communities are well preserved, making it known as the "green pearl" in central China. The park is rich in animal and plant resources, including 1461 species of wild vascular plants. Forexample, there are 82 species of ferns belong to 49 genera of 27 families, 22 species of gymnosperms belong to 16 genera of 8 families, and 1357 species of angiosperms belong to 598 genera of 160 families. Pinus wuqiushi of Dabie Mountain is listed as the first group of national rare tree species under Second-class protection, and the national second-class key protection wild plant. It is only distributed in the zhongshan zone at 700～1200 meters above sea level at the border of Anhui and Hubei provinces. Besides ferns, gymnosperms, the angiosperms of Dabieshan Mountain are also very prosperous, such as ginkgo, Southern Yew, ju tree and so on.

千年古树 Thousand-year old tree

苍劲古树 Vigorous tree
摄影：汪利群 Photographed by Wang Liqun

黄山松1 Huangshan pine No.1

黄山松2 Huangshan pine No.2

良田美景

/ Splendid view of fertile fields /

草滩过盘 Grass beach

百花争艳 / Flowers in Bloom /

樱桃 Cherry 摄影：华仁 Photographed by Hua Ren

蝶花之吻 Butterfly kiss the floor

杜鹃花 Rhododendron

桔梗花 Platycodon grandiflorum flower

野生杜鹃花 Wild rhododendron

红尾水鸲 Rhyacornis fuliginosus
摄影：周卫烈 Photographed by Zhou Weilie

中华秋沙鸭 Chinese merganser
摄影：周卫烈 Photographed by Zhou Weilie

灰头麦鸡 Grey head lapwing
摄影：周卫烈 Photographed by Zhou Weilie

鸳鸯 Mandarin duck
摄影:周卫烈 Photographed by Zhou Weilie

林麝 Forest Musk Deer

小灵猫 Small indian civet

猕猴 Macaque

摄影：华仁 Photographed by Hua Ren

奇人出奇山
Home of Celebrities

文化底蕴

/ Cultural Deposits /

“江山如画，一时多少英雄豪杰”。黄冈崇文重教，文化底蕴深厚。这里是著名的诗词之乡，苏轼的《念奴娇·赤壁怀古》曾被誉为“古今绝唱”，孕育了著名的东坡文化；这里红旗漫卷将星闪耀，发生了黄麻起义、刘邓大军千里跃进大别山等革命事件，积淀红色文化，传承红色基因；这里基础教育闻名遐迩，黄冈中学享誉海内外，是著名的教育之乡；这里还是历史悠久的宗教文化圣地，有著名的南武当道教文化、东山五祖寺的佛教文化、禅花自在香的禅宗文化。

"Picturesque rivers and mountains of our land! How many heroes are involved and how grand!" Huanggang values literature and pays attention to education, thus it has profound cultural deposits. It is the home of Chinese poetry Sushi's *Niannujiao ·Chibihuaigu* that was once known as "The Best Poem Both in Ancient and Modern Times", thus there gave birth to the famous Dongpo culture. It has red flags shining brightly and has given birth to a number of generals, since there has happened many revolutionary events, such as Huang'an-Macheng Uprising, and the Army led by Liu & Deng Leaping forward into Dabieshan. Through those events, they can help to accumulate red culture, and attribute to inheritance of red genes. It is well known for its basic education, and Huanggang Middle School enjoys a good reputation both at home and abroad. It is also a holy land of religious culture, including the famous Taoist culture of South Wudang, the Buddhist culture of Dongshan Wuzu Temple, and the Zen style of Flower fragrance.

历史名人

/ Historical Figures /

“千年英雄地，灵秀举世奇”。黄冈有着厚重的人文底蕴，四大发明之一的活字印刷术发明者毕昇，明代医圣李时珍，药圣万密斋，中国地质学家李四光，中国京剧鼻祖余三胜等1600多位历史名人都诞生于黄冈。黄冈有着辉煌的红色历史，是红色的摇篮，在这片土地上，诞生了3个中共一大代表、2位国家主席以及200多名开国将帅，“两百个将军，同一个故乡”写的就是黄冈故事。

"A heroic land of thousands of years, a beautiful scenery incomparable in the world." Huanggang has a rich cultural heritage. More than 1,600 historical celebrities were born in Huanggang City, including Bi shen, the inventor of typography, which is one of the four great inventions of China; Li Shizhen, the medical sage of Ming Dynasty & medicine sage Wan Mizhai, the Chinese geologist Li Siguang and Yu Sansheng, the founder of Chinese Peking Opera. Huanggang has a brilliant red history, the cradle of the red culture. On this land, there has born 3 representatives of the First National Congress of the Communist Party of China, 2 national presidents and more than 200 founding generals. "200 Generals, One Hometown" is the story of Huanggang.

京剧鼻祖——余三胜 The Originator of Peking Opera—Yu Sansheng

地质学家—— 李四光
Geologist—Li Siguang

活字印刷术发明者—— 毕昇
Inventor of typography—Bi Sheng

医圣—— 万密斋 Medical sage—Wan Mizhai

药圣—— 李时珍 Medicine Saint—Li shizhen

东坡赤壁 Dongpo Chibi (Red cliff)
来源于网络 https://image.baidu.com

东坡文化——赤壁浮雕墙
Dongpo culture—Relief wall of Chibi

赤壁浮雕墙 Dongpo Culture-Relief Wall of Chibi
背景为砂岩，主景为花岗岩
The background is composed of sandstone and the main landscape is granite

教育文化——黄冈中学
Education culture—Huanggang Middle School

红色文化——中国工农红军第四方面军指挥部
Red culture—The Fourth Front Army Headquarters of the Chinese Workers' and Peasants' Red Army

红色文化——革命老区 Red culture—Old revolutionary base area

戏曲文化 —— 黄梅戏 Opera culture—Huangmei opera

农耕文化 Farming culture

民俗工艺 / Folk Craft /

秀丽的山川和悠久的历史共同铸造了黄冈非物质文化和民俗特产，如英山的黄梅戏、罗田的特色美食——吊锅。

Beautiful mountains and long history both attributes to the intangible cultural heritage of Huanggang City and its specialties, such as Huangmei Opera in Yingshan County, and specialty cuisine—Hanging Pot in Luotian County.

民俗表演——舞龙 Folklore performance—Dragon dance
摄影：徐晖 Photographed by Xu Hui

民间工艺——红安绣活
Folk crafts—Hong'an Embroidery

民间工艺——英山缠花
Folk crafts—Yingshan entangled flowers

民间工艺——黄梅挑花
Folk crafts—Huangmei cross-stitch work

民间工艺——皮影戏 Folk craft—Chinese shadow puppetry

民间工艺——编竹篮 Folk craft—Weaving bamboo basket

特色产品——手工油面 Speciality—Handmade oil noodles
摄影：龙钢 Photographed by Long Gang

特色产品——英山茶叶 Speciality—Yinshan tea

板栗 Chestnut

特色产品——黄冈甜柿 Speciality—Huanggang sweet persimmon

罗田吊锅 Luotian hanging pot

糍粑 Glutinous rice cake

火烧粑 Huoshaoba

英山毕升饼 Yingshan Bisheng cake

深山藏古居

Ancient Homes Hidden in Remote Mountains

山中古宅——段氏府
Ancient residence in the mountain - Duan's

深山藏古居

/ Ancient Homes Hidden in Remote Mountains /

“滚滚长江水，苍苍大别山”。大别山，一座绵延八百里的古老山脉，山之美，在于巍峨高耸，挺拔险峻，峰峦重叠；在于奇峰秀水，绿林扬风，生机勃勃。

“古今人老尽，山水镇长闲”。山之美，也在于山水中的人间烟火气。在大别山的深山之中，有着神工天巧、风格迥异的各类建筑。一是神秘的道教建筑——南武当、古老佛教建筑——柏子塔、悠久历史建筑——东坡赤壁；二是充满红色印记的革命遗址，如董必武故居、李先念故居、红安长胜街等。除此之外，山中古街民居、传统古镇、传统村落、古色民宿与新时代的美丽乡村、美丽社区等现代化建筑交相辉映，构成一幅美丽画卷。

"The Yangtze River keeps flowing, while Dabieshan has green vitality." Dabieshan is an ancient mountain range stretching 800 li. The beauty of the mountain lies in their lofty heights, tall and steep ridges, and their overlapping peaks; the beauty of the mountain lies in extradinary peaks and clean water, green forest and mind wind, which is full of vitality.

"The people of the past and present will grow old, but the mountains water and towns will always be here". The beauty of mountains also lies in the human fireworks in the landscape. In the deep mountains of Dabie Mountain, there are all kinds of buildings with different styles. One is the mysterious Taoist architecture — South Wudang, ancient Buddhist architecture —Baizi tower, long historical architecture —Dongpo Chibi; The other is the revolutionary sites full of red marks, such as dong Biwu, Li Xiannian, Hongan Changsheng Street, etc. In addition, the ancient streets in the mountains, folk houses, traditional towns, traditional villages, ancient homestays, and modern buildings in the new era, such as beautiful villages, beautiful communities, and reflect each other, forming a beautiful picture.

宗教建筑——红安天台寺 Religious architecture—Hong'an Tiantai Temple

宗教建筑——五祖寺 Religious Architecture—Wuzu Temple

红色建筑——董必武故居 Red building—Dong Biwu's Former Residence

红色建筑——李先念故居 Red building—Li Xiannian's Former Residence

革命遗址——红安长胜街 Revolutionary site—Hong'an Changsheng Street

九龙寺 Jiulong Temple

传统村落——大河冲村 Traditional village—Dahechong Village

传统村落——左家沟村
Traditional village—Zuojiagou Village

新时代美丽乡村 New era beautiful countryside
摄影：方华国 Photographed by Fang Huaguo

西河之春 Spring of the West River
摄影：孔方乡 Photographed by Kong Fangxiang

古街民居——红安七里坪古镇 Ancient street residence—Qiliping ancient town, Hong'an

建设
路漫漫
The Journey with Promising
Future Ahead

公园建设

/ Construction /

公园网站 Geopark website

公园监控设施 Geopark monitoring facilities

公园栈桥 Geopark trestle

薄刀锋景区 Bodaofeng scenic area

摄影：华仁 Photographed by Hua Ren

天堂寨游客中心 Tiantangzhai visitor center

公园餐饮服务设施 Geopark catering service facilities

生态停车场 Ecological parking lot

地质环境监测站 Geological and environmental monitoring station

地质遗迹保护点 Geosites protection point

公园解说牌1 Geopark bulletin board No.1

公园解说牌2 Geopark bulletin board No.2

公园信息电子展示屏 Geopark electronic interpretation screen

公园界碑 Geopark boundary monument　摄影：华仁 Photographed by Hua Ren

公园警示牌 Geopark warning pannels

公园信息导向牌 Guide boards of geopark information

科普教育——大别山地质公园博物馆

/ Popular Science Education—Dabieshan Geopark Museum /

2017年5月15日，湖北黄冈大别山地质博物馆正式开馆，并作为中国地质博物馆分馆对外开放。湖北黄冈大别山地质博物馆馆藏丰富，馆内设立了地球的故事，中国的脊梁、巍巍大别山、大别山生态园、大别山名人馆和地质标本精华厅等8个陈列厅。博物馆具有很强的地质科普价值，深受国内外游客的喜爱。

On May 15, 2017, Dabieshan Geopark Museum was officially opened, and as a branch of the Chinese Geological Museum, it is open to the public. Dabieshan Geopark Museum has a rich collection of eight exhibition halls, including the Story of the Earth, the Backbone of China, the majestic Dabieshan, the Dabieshan Ecology Garden, the Celebrity Gallery of Dabieshan and the Geological Specimen Essence Hall. The museum has a strong geological science value and is loved by domestic and foreign visitors.

大别山地质公园博物馆 Dabieshan Geopark Museum

地质标本精华厅
Geological specimen essence hall

名人馆
Celebrity Gallery of Dabieshan

水晶石 Crystal stone

雄黄 Realgar

纹石 Veins stone

黄河象化石 Fossil of Stegodon huanghoensis

方解石萤石 Calcite and fluorite

自然铜 Pyritum

金色方解石闪锌矿
Gold calcite and sphalerite

方解石 Calcite

鹦鹉嘴龙化石
Fossil of psittacosaurus

外国友人参观博物馆 Foreigners visit the museum

举办世界清洁地球日活动 Clean-up day activities

博物馆科普走廊 Popular Science Corridor in the Museum

博物馆内精美藏品1
Exquisite collections in the museum No.1

博物馆内精美藏品2
Exquisite collections in the museum No. 2

博物馆内精美藏品3
Exquisite collections in the museum No. 3

科普教育——教学实习基地

/ Popular Science Education—Teaching Practice Base /

野外研学教育
Field education

展示讲解
Presentation and explanation

科普进校园

Popular science into campus

科普进社区

Popular science into the community

交流学习——访问巴西阿拉里皮世界地质公园

Exchange learning—Visit the Aralipi World Geopark in Brazil

科普教育系列丛书

/ Series of Books for Popular Science Education /

社区发展 / Community Development /

地质公园的建设离不开地质公园社区居民的支持。在大别山区，有麻城市朝圣门社区、英山县周家畈社区、罗田县外婆桥社区和十人寨社区等优秀地质社区。当地社区居民不仅是地质公园最直接的受益者，也是地质公园的守护者、建设者和导游员。

The establishment and development of a Geopark should be based on strong community support and local involvement. In Dabieshan Mountain area, there are excellent geopark communities such as Chaoshengmen Community in Macheng City, Zhoujiafan Community in Yingshan County, Jiuzihe Town Community in Luotian County and Shirenzhai Community. Local community residents are not only the most direct beneficiaries of the geopark, but also the guardians, builders and tour guides of the geopark.

十人寨地质社区 Shirenzhai geocommunity
摄影：林涛 Photographed by Lin Tao

社区建设——丽景山庄 Community construction—Lijing Villa

西河十八湾 Villages in Xihe River

新时代社区广场——余三胜广场 New Times Community Plaza—Sansheng Plaza
摄影：方华国 Photographed by Fang Huaguo

英山茶园
YingShan tea garden

社区产业——农耕产业
Community industry—Agricultural industry

社区产业——光伏发电
Community industry—Photovoltaic power generation

节事活动 / Festival Activities /

黄冈大别山世界地质公园开展了各式各样、针对不同群体的科普教育活动，加大地质公园的宣传力度，举办了“自行车大赛”“英山茶叶节”“罗田红叶节”和“罗田吊锅美食节”等民事活动，扩大周边社区的参与，推动了地方经济的可持续发展。

Huanggang Dabieshan UNESCO Global Geopark has carried out a variety of popular science education activities targeting for difffferent groups. In order to increase the publicity of geopark, civil activities such as "Bicycle Race" "Yingshan Tea Festival" "Luotian Red Leaf Festival" and "Luotian Hanging Pot Food Festival" have been held to expand the participation of surrounding communities and to promote sustainable development of local economy.

峡谷漂流节 Canyon Rafting Festival

南武当滑雪场 Nanwudang Ski Resort
摄影：张新安 Photographed by Zhang Xinan

瀑布降索
Waterfall Descent Rope

自行车大赛
Bicycle Race

英山茶叶节
Yingshan Tea Festival

罗田甜柿节
Luotian Sweet Persimmon Festival

2020麻城首届乡村文化旅游节
The first Macheng Rural Cultural Tourism Festival (2020)

罗田吊锅美食节 Luotian Hanging Pot Food Festival
摄影：陈永斌 Photographed by Chen Yongbin

大别山麻城民俗节
Dabieshan Macheng Folklore Festival

天堂寨第七届天贶民俗文化旅游节
The 7th Folk Culture Tourism Festival in Tiantangzhai
来源:天堂寨旅游景区官网
From the official website of Tiantangzhai Scenic Area

图书在版编目(CIP)数据

行走在黄冈大别山世界地质公园/李江风等编；张地珂,李凤，杜冬琴译.—武汉：中国地质大学出版社，2023.5

ISBN 978-7-5625-5595-7

Ⅰ.①行… Ⅱ.①李… ②张… ③李… ④杜… Ⅲ.①大别山-地质-国家公园-地质-黄冈 Ⅳ.①S759.93

中国国家版本馆CIP数据核字(2023)第106544号

行走在黄冈大别山世界地质公园

李江风 彭文胜 黎 妍 邓志高 高志峰 吴 笛 编

张地珂 李 凤 杜冬琴 译

责任编辑：舒立霞　　责任校对：何澍语

出版发行：中国地质大学出版社（武汉市洪山区鲁磨路388号）　　邮编：430074

电话：（027）67883511　　传真：（027）67883580　　E-mail：cbb@cug.edu.cn

经销：全国新华书店　　http：//cugp.cug.edu.cn

开本：880mm×1230mm 1/12　　字数：450千字　　印张：14

版次：2023年5月第1版　　印次：2023年5月第1次印刷

印刷：湖北睿智印务有限公司

ISBN 978-7-5625-5595-7　　定价：178.00元